Typesetting by Deanta Global Publishing Services, Chennai, India

CW01584103

Edition: 1
Version: 1

WORLD LAND TRUST™
www.carbonbalancedprint.com
CBP2243

Corporate member of Plain English Campaign
Committed to clearer communication
557

QUICK REFERENCE GUIDE TO RISK PROFILING

NEBOSH

nebosh

Contents

Introduction

Who is this guide for?

This guide is intended as a quick reference resource for anyone involved in the risk profiling process in their organisation, including health and safety professionals.

In health and safety, the focus is often on the threats to be treated, but it is also essential to consider the flip side – the opportunities that managing risk brings. The range of considerations for a risk profile goes beyond health and safety, encompassing both internal and external sources and consequences of risks. Producing a risk profile for an organisation allows leaders to prioritise and treat risks (both positive and negative) and is a crucial tool for business planning. While it is important to acknowledge that 'risk' is not a purely negative thing, the focus should be on reducing the likelihood and severity of a negative outcome.

This guide outlines the key purpose of the risk profiling process and the critical areas of consideration when assessing risks to an organisation. It also introduces some established tools and models that can help identify risks and opportunities. It is intended as an informational tool only. It is not intended as a training resource to support the study of NEBOSH qualifications.

About NEBOSH

The National Examination Board in Occupational Safety and Health (NEBOSH) is a leading global organisation which provides health, safety, environmental, risk and wellbeing qualifications. Our internationally recognised qualifications help to raise the competence of safety and environmental professionals as well as individuals at all levels in the workplace. NEBOSH qualifications are highly respected by governments, employers and our learners.

Find out more at nebosh.org.uk

Key Definitions

Threat is something that could cause damage or danger or make an organisation unable to achieve its objectives.

Opportunity is something that can benefit an organisation or its objectives.

Risk is the effect of uncertainty on objectives. Risk can be positive (opportunity) or negative (threat).

Consequence is the outcome of the threat being realised. Consequences can be positive as well as negative.

Risk profiling looks at the nature and levels of threats faced by an organisation. It examines the likelihood of adverse effects occurring, the level of disruption and costs associated with each type of risk and the effectiveness of the control measures in place.[1]

Risk register is a record of information about identified risks[2] (sometimes referred to as a risk log).

Control measures are actions taken or measures put in place to prevent or mitigate risk or consequence. The term **Risk treatment** can also be used to describe the action(s) taken once a risk has been identified and evaluated.

Notes

1. HSE, Managing for health and safety (HSG65, 3rd edition, 2013) (www.hse.gov.uk).
2. BS 31100:2008 Risk Management. Code of Practice. Glossary.

What is risk profiling?

Main aims

The purpose of risk profiling is to make risks visible within an organisation. At its simplest, a risk profile is a description of a set of risks. The profile can relate to the whole organisation or at a departmental level.

An organisation's risk profile should inform all aspects of the approach to leading and managing health and safety risks. Health and safety leaders need to ensure that their respective organisations have built a risk profile that covers:

- The nature and level of the threats faced by an organisation
- The likelihood of adverse effects occurring
- The level of disruption and costs associated with each type of risk
- The effectiveness of controls in place to manage those risks

The outcome of risk profiling will be that the right risks have been identified and prioritised for action, controls communicated, with minor risks given lower priority. A risk profile also informs decisions about what risk control measures are needed and the availability and allocation of resources.

Every organisation will have its own risk profile, and effective leaders must know the risks their organisations face, rank them in order of significance and take action to control them. This is the starting point for determining the greatest potential issues for an organisation.

TIP

Clarifying risks at the strategic level can help those who write and update risk assessments. This could lead to additional benefits, such as encouraging knowledge sharing and a consistent approach within the organisation.

FOCUS

Difference between risk profiling and risk assessment

Risk profiling involves assessing risks. However, in health and safety, the term 'risk assessment' usually refers to a slightly different process, often focused on the risk arising from a specific workplace activity or process. The hazards, risks, and controls it considers tend to be more direct.

Risk profiling is a more high-level activity carried out at the business unit, division or branch level. Its output is intended to identify areas of high risk or undesirable exposure to risk that require senior leadership attention. These threats are considered in terms of the potential harm to people, disruption to the organisation and financial impacts, in order to determine where resources should be prioritised effectively.

So, a risk profile is a form of high-level risk assessment, but with the distinguishing feature that risk profiling considers and consolidates a range of threats and a range of risks, not just those related to health and safety.

Outputs

The outputs of the risk profiling process will be a prioritised register, enabling organisations to focus on the most significant issues and not give unnecessary priority and resource to minor risks. It also outlines what risk control measures are already in place to inform decisions about what more may be needed.

A risk register is a tool an organisation can use to record and monitor risks identified through the profiling process. It provides an overview and prioritisation of risk for management attention. It usually describes each risk, an assessment of the likelihood and consequences, a ranking, a risk owner and mitigation information. When populated with information on each risk, the register can be analysed to give risk profiles for different aspects of the organisation. Over time, it can be used to help recognise trends in the profile and to focus management attention on the areas of greatest risk.

EXAMPLE
Risk register template

Ref	Category	Description of risk	Description of impact/ opportunity	Current mitigation	Likelihood (low, medium, high)	Level of impact (low, medium, high)	Recommended treatment	Risk owner

Risk profiles are not static and require review at intervals, especially when significant change occurs (for example, the introduction of new processes or an acquisition), and should be incorporated into a management of change process. In some situations, the risk profile can change throughout the lifetime of a project, for example, during a major construction project.

TIP

Ensure your risk profile table is similar across other disciplines so combining is easy and may allow particular risks to be prioritised and/or highlight opportunities. Combining registers can help with integrated management systems and help auditors evidence strong organisational risk management.

Who should be involved in risk profiling?

Competent persons with the appropriate skills and experience relevant to the organisation should conduct risk profiling. These persons will likely be a cross-functional group from different levels of the organisation.

Leadership

An important role of leaders is to take ownership of health and safety and other risks. This leader might be an organisation's chief executive or owner, a specially constituted committee or group, or a senior manager.

There are many reasons why health and safety risks need to be considered as part of a high-level process:

- **Commercial, operational and safety goals can be optimised when considered together**. Considering different types of business goals together should identify critical safety requirements early in the process. This allows the most effective risk reduction options to be chosen, and risks can be designed out wherever possible. Identifying these requirements later in the process often means that earlier choices must be reversed, leading to delays and unbudgeted costs. Sometimes it is impractical to reverse earlier choices, forcing decision-makers to select from poor options.
- **Safety connects with other business interests**. Many organisations have closely coupled systems and procedures. For example, there are usually multiple

facets to safety risks with connections to finance, environment, reputation, compliance and security. Making a change to one process, procedure, or activity can have knock-on effects elsewhere in the organisation. It is essential that evaluation and planning are done with the input of people who have sight across the entire organisation or department.

Leaders will need to keep an overview of the risk profiling process and be aware of the major risks within the organisation. They should also ensure that minor risks have not been over-prioritised.

Risk profiling should be a key part of developing and reviewing any business plan, so the two should be considered together, and the same people should be involved in both activities.

Cross-functional input

Any risk profile will only be as good as the information gathered. It is important to consider this at the outset. It is essential that those involved in creating the risk profile are competent to do so and can provide the necessary depth and breadth of information on the organisation's operations and associated risks and controls.

People involved in different areas of operation are also best placed to foresee the potential impacts and consequences that disruption might bring. This approach can also mean consulting with workers at all levels – not just senior management.

The process provides transparency when all organisational risks are presented using the same units of measurement for all activities, all areas and all consequences. In turn, this assists the organisation-wide decision-making process for continuously reducing risk.

Managers

Managers are well placed to support the identification of hazards and threats, the evaluation of risks and the selection and implementation of control measures. Management will also have insight into what risks should be prioritised over others.

In addition to producing the risk register, managers will need to give extra consideration to worker consultation and involvement. In practice, this means communicating with worker representatives, explaining the risk profile and ensuring that everyone has the necessary information, instruction and training to appreciate the risks identified. Workers also need sufficient training and information to understand their role in operating necessary risk control measures, and managers play a vital role in this ongoing process.

Health and safety professionals

It is likely that health and safety professionals will lead or have key responsibilities during the risk profiling process due to their skillset and experience in risk management and control. Major hazard organisations may require specialist external help.

The risk profiling process

How risk profiling is carried out can vary across sectors. Making it a collective exercise involving relevant people from across the organisation, unit or discipline for which the risk profile is being produced is always good practice.

Risk profiling usually involves the following key stages:

- Agree definitions and descriptors for types of risk, types of consequence, likelihood, treatment and controls, and criteria for risk prioritisation.
- Select the risk profiling team and gather information about operations and processes.

- Understand the context and risk appetite of the organisation.
- Identify potential risks, considering both external and internal causes.
- Understand the potential consequences if those risks are realised.
- Consider existing and additional controls.
- Agree prioritisation of risks.

- Produce a risk register to record findings and inform further actions.
- Ensure continuous review and improvement.

Organisational context and risk appetite

All organisations, whether large corporate entities, small organisations, individual departments, teams, or projects, operate in a context. The risk profiling process needs to start with an understanding of the external and internal scope and context of the organisation, department or function. This understanding assists with focusing the process on what matters most and helps to ensure that risks are assessed and treated effectively.

The primary purpose of setting out the scope and context is to ensure that all of the organisation's aims, objectives, purpose, core processes and activities are considered along with the organisation's external environment. Scope can be defined at different levels, including strategic, operational and programme/project.

The starting point for scope and context must always, therefore, be full consideration of the organisation's:

- Vision, objectives, and strategic plan
- Culture and values
- Structure and design
- Capital, skills, and other resources

Consideration of the types and areas where risks or weaknesses might be present should be included, for example:

- IT or telecommunications
- Product quality

- Financial risks
- Key competence requirements, eg, lack of talent or key skills
- Health, safety, or environmental incidents

It is also crucial to understand the risk appetite of the organisation. Some organisations are 'aggressive' and keen to expose themselves to certain types of risk if there is the potential for the reward to be greater. Other organisations are much more cautious and are either less likely to engage in 'high-risk' strategies or they are quicker to react to potential threats when they become visible. It is likely that an established organisation will already have a clear picture of their risk appetite, and this should be referred to throughout the risk profiling process.

Types of risk

Risk identification lies at the heart of the risk profiling process. If risks are realised, they will affect the organisation's set objectives. A good starting point for risk identification is to look at the organisation's business, operational, or project plan and analyse the core processes. It is important to remember that a risk profile outlines a set of risks – not just health and safety ones.

FOCUS

Approaching the identification of risk categories

When identifying risks, think about all the risk categories that affect the organisation, not just the obvious ones or those featuring most prominently in the media at the time. One approach is to look at risks 'top down', 'bottom up', and 'across' the organisation, function, or project. This approach can be invaluable in helping to ensure that everything is considered. Think about all the things the organisation relies on being in place (in other words, its dependencies) to achieve its objectives and recognise any key constraints to achieving them. Sometimes these risks could be so significant that it might be necessary to change plans to avoid or reduce the level of risk faced.

Examples of risk categories are:

- Personal injury
- Ill-health
- Legal/compliance
- Equipment

- Fire
- Security
- Finance
- Business (strategic or operational)
- Reputation
- Environment
- Quality
- Technology

EXAMPLE

Identifying different risk categories

An intriguing example of identifying different risk categories was highlighted during an organisation's risk profiling process. The first risk category identified related to the threat of health and safety risks faced by forklift truck drivers. The second category involved a quality and technology risk arising from competitors using driverless forklift trucks.

This assessment identified an opportunity to simultaneously treat the threat of safety risks for drivers (eg, collisions and injuries) and gain a technological advantage over competitors. As a result, the risk register highlighted the potential to improve driver safety while utilising advanced technology to outperform industry rivals.

This example shows the importance of identifying and addressing various risk categories, thereby maximising opportunities for improvement and innovation within an organisation.

TIP

Risk categories can have codes to make referencing and updating easier. For example, noise risks could be classed as HSN (Health Safety Noise). Machines may then be given the code 'HSN1 – Machine Noise'. This reference can then be used in risk registers and other documentation. The benefit of doing this is that it is easy to identify and locate all HSN1 risks and make the necessary changes when there is an update.

Good risk identification should identify and consider:

- Relevant, appropriate, valid, and up-to-date information – good risk identification concentrates on risks that can affect the organisation's objectives and is based on consideration of experience, current and future contexts
- Tangible and intangible sources of risk – not all risks can be measured or quantified. Just because a risk cannot be calculated, that does not mean that it should be excluded
- Previous events can also be a useful guide to the future. Always look at the 'root causes' of previous incidents
- Causes and consequences
- Threats and opportunities
- Changes in external and internal scope and context
- Indicators of emerging risks – what might happen within the organisation, industry, or sector, and how those changes might create opportunities and threats

- Nature and value of assets and resources – the nature and size of the organisation and the nature and size of the risk.
- Potential consequences and impact on objectives
- Levels of existing control

External factors

Multiple external factors affect how organisations can operate and many of these can be unpredictable. Still, it is important to be prepared for potential scenarios and to adapt as global industries shift.

Some commonly recognised external factors to consider are:

- Acquisition or merger
- Changes to banking or other financial instruments
- Changes in the balance between primary (extraction) industries, secondary (manufacturing and production) industries, and service industries in different parts of the world
- Regulation (or de-regulation) of industries by governments
- Changes in technology can affect the nature of many industries and the consequent demand for products and services
- Legal changes
- Changes in socially acceptable behaviours and expectations
- Changing customer expectations or requirements

Finally, it is important to consider any organisation in a global context, not just in the country of operation.

Internal factors

Factors internal to an organisation are crucial to understand what risks may be present, as well as to understand potential consequences. Understanding the organisation is also key in determining how treatments, controls, and mitigations can be applied.

Internal factors relevant to most organisations include:

- Purpose, objectives, strategies and plans
- Culture, behaviours and values
- Organisation structure and design
- Systems and processes
- Availability and accessibility of resources such as capital, materials, people and skills

Throughout the risk profiling process, it is good practice to revisit these internal factors. If one department makes a change to one of its processes, procedures or activities (as a result of a risk being realised or as part of a control measure), there are likely to be implications for other adjoining departments or activities.

It is also useful to ensure that existing internal documents, such as legal registers and the organisation's health and safety policy, are considered to support risk identification and prioritisation.

Stakeholders

Another way to consider risks to an organisation is to consider those groups of people with an interest or 'stake' in the business. These people are called stakeholders, sometimes called interested parties, and can be internal or external to an organisation.

Some common categories of stakeholders are:

- Shareholders
- Employees
- Financial institutions (eg, lenders)
- Regulators
- Trade unions
- Trade association
- Media
- Local government
- National government
- Suppliers
- Competitors
- Customers
- Local community

This list includes stakeholders directly linked to the organisation as it is currently. A more complex stakeholder analysis might consider potential customers or the effect of dominant customers on others. A stakeholder model could be presented in a way that details different risks associated with different groups or individuals or identify potential stakeholders who would only become relevant if the risk was realised.

TIP

Surveying these stakeholders to ask what risks and opportunities they see can be an excellent way to receive good information and critically evaluate your risk profiling process and registers.

Tools and models

There are a variety of tools and techniques that can support risk identification as part of the risk profiling process. These include risk workshops, brainstorming, interviews, checklists, root cause analysis and scenario analysis.

Porter's Five Forces Model

In 1979, American academic Michael Porter published a five-element model to help organisations to consider how much power and control they have over the environment in which they operate.

The five elements are:

Figure 1 Porter's five elements

Competition and competitiveness

In a market where an organisation's products or services are unique, and there is no competition, they have discretion over price, availability, etc. If they are one of several organisations competing for the same business, they are likely to have much less control.

Market accessibility to new entrants

If an organisation offers something unique and original, or their product requires a high level of capital investment in equipment or infrastructure, it is much harder for new entrants to get to this market so easily and provides the organisation with greater control.

Supplier power

If the market that an organisation operates in is highly dependent on the cost and availability of raw materials or services, then its suppliers have much greater control over it than they do. The suppliers' control over supply and charges will affect the organisation's product availability and prices.

Purchaser power

If the customer dominates the market that an organisation operates in (eg, farmers are heavily reliant on supermarkets), then the organisation is less likely to have discretion over supply or prices.

Threat of substitution

If an organisation's product can be easily substituted for an alternative, there will always be limits to the extent that they can influence their market.

EXAMPLE

Using Porter's Five Forces model

This example uses Porter's Five Forces model to explore risk profiling within the context of the traditional taxicab industry.

Competition and competitiveness: This is a highly competitive industry. In addition to competing with other taxi companies, traditional businesses must now contend with digital and mobile app ride-hailing platforms that have disrupted the market. The risk associated with industry rivalry includes price wars, aggressive marketing strategies, and constant innovation.

Market accessibility to new entrants: The industry faces a moderate threat of new entrants due to the low barriers to entry. The risk associated with new entrants lies in their ability to attract customers with lower prices, improved convenience, and advanced technology.

Supplier power: Suppliers have moderate bargaining power. The industry relies on vehicle manufacturers, fuel providers, and maintenance and repair services. Rising fuel and energy costs, vehicle prices, or repair charges can decrease profit margins and increase operating expenses.

Purchaser power: Customers possess moderate bargaining power, given the availability of multiple alternatives. They have the ability to choose services based on factors such as price, availability, convenience, reliability, and service quality.

Threat of substitution: The industry faces a high threat of substitutes, primarily from digital mobile app ride-hailing services. These services offer similar transportation solutions but with added convenience, real-time tracking, and flexible pricing. The widespread adoption of smartphones and the ease of use of ride-hailing apps have significantly impacted the demand for traditional taxis.

STEEPLE model

The STEEPLE model is useful in helping to identify external and internal risks that create threats and opportunities for your organisation.

Each letter represents a category of factors to consider:

Table 1 STEEPLE model

S	Social	Age, health, mortality, education, media, attitudes to crime and criminals, culture. People remaining in work for longer and multi-generational workforces are an example of social risks.
T	Technological	Technological developments, rate of change, investment in development, impact of new technology. An example is the rapid development of AI and its potential use in many sectors.
E	Economical	National and international economic factors, inflation rates, spending power, interest rates, exchange rates, poverty. Factors such as acquisitions and mergers can also be included in this category.
E	Environmental	Geographic factors and limitations (where the organisation is based, where its customers are based), the terrain, access to services and utilities (such as network connectivity), and the climate (temperature, rainfall, etc). It can also include risk from natural disasters such as flooding.

P	Political	Political stability, government policy, the role of local or national elected officials. Policies on taxation, employment or environmental law, health, and safety and attitude to regulation and enforcement.
L	Legal	The extent of law – what organisations can and cannot do, and the extent and level of sanctions. Organisations may base themselves in some countries because they know it is easier to operate there.
E	Ethical	Increasingly organisations are expected to address ethical risks, largely those which reflect the expectations of their customers. Issues such as environmental damage, gender equality, workers' rights, and cultural sensitivities can be included in this category.

Thinking about an organisation using the STEEPLE model can help to identify threats and opportunities that might not have been previously considered. The challenge with this model is to think beyond the uncertainty and identify the risk (opportunities and threats) that uncertainty creates for the organisation. A good way to think of the model is to spot an uncertainty (like climate change) and convert that into a business risk (like supply chain failure).

SWOT Model

A SWOT analysis continues the evaluation of internal and external factors, by identifying which of these are strengths and weaknesses of an organisation, as well as potential opportunities and threats.

Table 2 SWOT analysis

S STRENGTHS	W WEAKNESSES
• Advantages • Capabilities • Resources	• Disadvantages • Vulnerabilities • Limitations
O OPPORTUNITIES	**T THREATS**
• Chances • Developments • Benefits	• Obstacles • External effects • Risks

Using the STEEPLE or Five Forces model will give a better understanding of the external and internal factors that are likely to affect an organisation – the Opportunities and Threats. An organisation's ability to respond to opportunities and threats is determined by its Strengths and Weaknesses.

Strengths are the things that an organisation does well. These can be a source of success, achievement or control.

Weaknesses are the things that an organisation might not be very good at. These can expose organisations to failure, poor performance, or lack of control.

British Standards Institution identified a series of 16 factors across the themes of Leadership, People, Process, and Product - they referred to these as the Organizational Resilience Index. These themes can be used to identify strengths and weaknesses:

Table 3 Sixteen factors comprising the Organizational Resilience Index

Leadership	People
Culture Community engagement Awareness, training, and testing Alignment	Leadership Vision and purpose Reputational risk Financial management Resource management
Process	**Product**
Governance and accountability Business continuity Supply chain Information and knowledge	Horizon scanning Innovation Adaptive capacity

Further information on the Organizational Resilience Index can be found on the BSI Group website: www.bsigroup.com/en-IL/Our-services/Organizational-Resilience/

TIP

Working on ways to improve an organisation's strengths and weaknesses may fall outside the scope of the main risk profiling activity. However, continual improvement in strength across any of the areas listed here will increase the organisation's ability to respond to any threats or opportunities that may arise in future.

'Power and influence' model

In this model, organisations consider stakeholders in the context of their power and influence. This is because organisations are unique, as are their interactions with the local community in which they operate. The model identifies each stakeholder and plots them in a grid based on their interests and power. This model guides how to manage and communicate with all stakeholders:

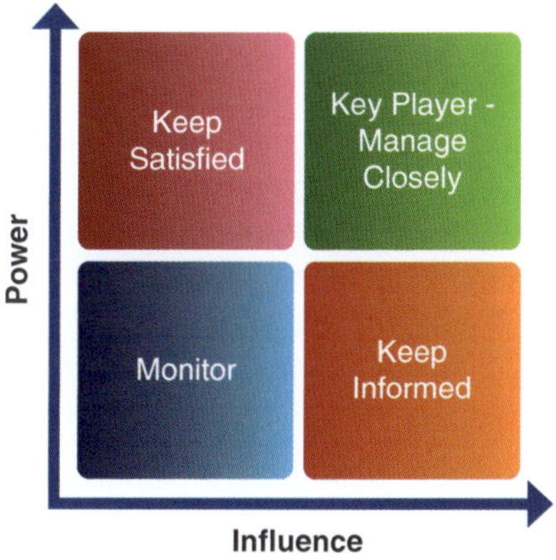

Figure 2 Power and influence grid

Potential consequences

Once potential risks are identified, it is crucial to think ahead and envisage the range of possible consequences that each risk could lead to. Again, this means thinking about the health and safety of people involved or affected and other types of harm that could come to the organisation.

Types of consequence include:

- **Personal injury:** a wide range of possible consequences, from minor injuries requiring first aid to fatalities.
- **Asset:** refers to things owned by the organisation and their high replacement value. Loss of or damage to equipment, machinery, vehicles, or premises needed by the organisation could have serious consequences for productivity, the ability to obtain future affordable insurance and reputation.
- **Security:** consider physical threats to site security, as well as cyber security.
- **Environment:** specific pollution incidents and the overall environmental impact of the organisation's activities should be considered when producing a risk profile.
- **Reputation:** this can range from individuals with complaints to sustained negative publicity in the industry or public press. It can also refer to internal reputation (eg, if it is widely known that the department dealing with contracts is very slow).

In addition to the type of consequence, consider the scale of impact. Generally, these are categorised as negligible, minor, serious, major, and catastrophic. Understanding the severity of potential consequences will help inform the prioritisation of risks and the level of proportionate controls that should be applied.

TIP

Recognising potential disruptions can give companies a competitive advantage. So, be sure to practice your emergency response to some of the more serious threats your risk profiling highlights.

Mitigation and controls

Once risks are identified, those involved in the risk profiling process need to confirm what control measures are already in place to address each risk. This allows the organisation to determine the level of risk that remains and recommend further controls where necessary.

FOCUS

Reasonable and proportionate

It is important to remember that risk management is not about completely removing all risks but ensuring that a reasonable and proportionate approach is taken to protect people, and the organisation, from harm. Risk appetite will be a strong determiner of which risks an organisation is prepared to tolerate and which risks need to be treated.

Properly considering risks and the possible eventualities of things going wrong allows controls to be implemented and work to progress. Often this allows for activities that may have been considered high risk to go ahead in a managed and sensible way.

Factors to be considered in selecting optimum risk controls must cover societal values, the organisation's values, risk-based analysis, technical judgement, good practice, and codes, standards and regulations. Be realistic about what is viable for the organisation at the current time.

Risk control strategies may be classified into four main areas:

- Risk avoidance
- Risk retention
- Risk transfer
- Risk reduction

Risk avoidance

Risk avoidance means the organisation consciously decides to avoid a particular risk altogether. This could be by discontinuing a process, procedure or activity.

Risk reduction

Risk reduction is a method for managing risks that attempts to reduce **likelihood**, **consequences** or both.

For example, reducing the **likelihood** of risk exposure in a car park can be achieved by limiting the number of interactions between vehicles and pedestrians. Options are:

- Reducing the speed limit
- Introducing pedestrian crossings
- Introducing walkways
- Operating one-way systems

A good example of reducing **consequences** is the wearing of personal protective equipment. The hazardous properties of the agent remain unchanged, but exposure to the risk has reduced consequences. So, exposure to silica dust when drilling into concrete walls is reduced by dust masks which are properly face-fitting.

Risk retention

This strategy retains risk within the organisation, which finances any consequent loss. Sometimes, this is referred to as being self-insured. Two options are possible: risk retention with knowledge and risk retention without knowledge.

Risk retention with knowledge

This strategy is where a conscious decision is made to meet any resulting loss from the organisation's financial resources. An example would be a public organisation with a large fleet of vehicles that decides it is better to pay reduced insurance premiums but to have higher voluntary excesses. Decisions on whether to retain risks can only be made once all the available information and trends are known.

Risk retention without knowledge

Risk retention without knowledge is where a risk exists, but the organisation omits to insure against it. The terms of an insurance policy may exclude a particular risk, but the organisation may be unaware of this term. For example, if an uninsured driver hits a vehicle, some insurance policies may not cover the cost of the damage.

Risk transfer

Risk transfer refers to two concepts. Firstly, it is the legal assignment of the costs of certain potential losses from one party to another, usually by an insurance policy. The terms of an insurance policy form the basis of a contract where the insurer undertakes to compensate the insured party for losses resulting from specified events.

Secondly, it is the transfer of activities to a contractor or specialist, rather than a direct employee of the organisation. This has the advantage – particularly with infrequent or technical tasks – of the organisation not having to invest in specialist equipment or training for jobs that rarely happen. An example would be where the activity is wholly outside of the organisation's expertise. For instance, constructing a new office would not be within the abilities of a waste recycling company. Instead, the task is subcontracted to a supplier specialising in constructing the required type of building.

Prioritisation

The final, key part of the risk profiling process is prioritising risks. This will help indicate which risks have the most potential to cause harm and which will require the most urgent attention. This then informs the order for controls to be implemented.

There are many approaches to recording prioritisation. Some organisations decide to order the register according to priority. Others might designate a level for each risk (ie, high, medium, and low). Remember that the priority level should be decided based on the current controls that are already in place. As further controls are implemented or as other factors change, the level of risk will change, and the register should be updated to reflect this.

Organisations can base their prioritisation on a range of factors:

Table 4 Factors to consider when prioritising risks

Factors to consider:	Sources of data to illustrate these include:
Societal values Taking account of the views, concerns and perceptions of society at large – which will include stakeholders. Examples might be environmental – influencing travel policy, or ethical - considering where raw materials are sourced.	News items, social media trends, and the views of focus groups, directors, employees and unions.

At certain times, such as war or recession, societal values can shift radically and may influence which activities are continued and which are not.	
Organisation values These are the values of the organisation. Company values take account of the views, concerns and perceptions of the stakeholders directly affected by the decision/option.	Published 'purpose, vision and values' of the organisation, history (previous decisions made by the organisation in similar circumstances), advertising and publicity materials that illustrate the organisation's values
Risk-based analysis Certain risk decisions lend themselves to risk-based analysis. This is largely determined by the availability of data and an understanding of the process being studied. For example, noise and vibration exposure can be analysed and used as a basis for further decisions about the protection that is needed to bring exposure to acceptable levels.	Qualitative surveys, exposure monitoring results and historic company data.
Technical judgement Expert judgement includes recognising what is good, best and emerging practice and applying sound engineering and scientific knowledge.	Use of experts (internal and external), reference to technical sources, academic and less formal reports, white papers and data sheets.

Good practice Good practice may include solutions that have yet to find their way into codes and standards. What is considered good practice may differ from situation to situation. Care should be taken to benchmark against the relevant good practice or emerging practice.	New or emerging technologies identified through personal contacts, trade associations, trade fairs, etc.
Codes, standards and regulations Codes, standards and regulations embody the lessons learned over the past years. Well-understood hazards and previous situations often provide an appropriate solution.	Legal registers to maintain contemporary banks of standards, regulations and codes.

These factors become especially important when there is a high degree of uncertainty with the risk. Organisations tend to gravitate towards their values and societal values when considering uncertainty.

TIP

Keeping evidence of risk prioritisation and decision making will help future risk register and profiling reviews. It is also good practice to record the process for audit purposes.

Recording information in a risk register

Once all information has been compiled, considered, and prioritised, it can be recorded in a risk register. This document can be circulated to anyone who needs to know about the risk profile of the organisation, or those who have responsibilities relating to mitigations.

EXAMPLE
Risk register for a taxi firm

Ref	Category	Description of risk	Description of impact/ opportunity	Current mitigation	Likelihood (low, medium, high)	Level of impact (low, medium, high)	Recommended treatment	Risk owner
BS01	Market share	Competition from app-based taxi services; Private customers (especially younger customers) prefer to use app-based booking system	Loss of business / loss of revenue	Past financial year focus on building relationships with corporate customers however these are not guaranteed	High	High	Join an app-based booking service which exists for local firms	Business manager

BS02	Market share	Increased competition from other local firms; This risk has existed for many years with little/ no change	Loss of business / loss of revenue	None. All local firms are similar size so competition is balanced	Low	High	None	Business manager
BO01	Workforce	Ageing workforce; Majority of our drivers have been with us for 10+ years and are looking to retire or reduce hours, also less willing to work night shifts	Loss of expertise/ knowledge through retirement; Loss of revenue if workforce is depleted	None	High	Medium	Consider recruitment strategy to reach younger drivers	Business manager

BO03	Equipment	All vehicles required to pass twice-annual inspection (Legal/License requirement)	Failed inspection results in costs to repair, additional business fine after 3 failed inspections across whole fleet	Drivers required to ensure annual service and regular vehicle maintenance	Medium	Medium	Provide clear information for drivers about their responsibilities; Consider contracting local mechanic for regular or pre-inspection checks	Shift manager
HS01	Safety	Lone working poses risk to our drivers' safety	Potential harm, injury or distress to our drivers	Policy of the drivers' right to refuse a fare. GPS tracking in vehicles	Medium	Medium	Consider CCTV in vehicles; Encourage drivers to report near-misses to better understand potential risk	Shift manager

Continuous review and improvement

Like a business plan, a risk profile should be a living document subject to regular updates and reviews. Once the risk register exists, setting a sensible regular review date is essential. This date will typically align with the highest residual risk calculated and depend on recommended control improvement actions. For example, if the highest priority risk identified in the risk profile is a moderate risk with additional controls planned for implementation within three months, it follows that the review date would be set for three months.

There are other reasons why it would be appropriate to review a risk register including:

Following an incident

Unfortunately, despite people's best efforts, things sometimes get missed, or the unexpected happens. In such cases, it is vital to investigate what went wrong and then take steps to prevent it from happening again. This may require revisiting the risk profile to add further risks, consequences or control measures.

Changes in processes or equipment

This includes technological changes that may not have been considered in the original review. The register should be updated to reflect any changes to how the organisation operates or how controls can be improved. Major changes in a work process may alter the prioritisation of a particular risk.

Changes in workforce

This may mean a change in the workplace's level of skills or competencies, which can influence the potential for accidents. This should lead to a reassessment considering the different skill and knowledge levels. It is crucial to ensure that responsibilities for control measures are reallocated where necessary.

Regular scheduled review

Over time the accuracy of the register may diminish due to changes in an organisation's priorities or changes to any of the external factors previously considered. For these reasons, it is important that the risk profiling exercise is regularly carried out to confirm that it is still relevant and applicable and that controls are still suitable.